fire

Technically Earth, Air, Fire and Water are not considered
elements in modern science. The Greeks included these
four as "elements" in their science because they believed
them to be the four basic qualities which made up all
substances.

Translation: W. Brian Altano

Pedagogical text: Cecilia Hernández de Lorenzo

First English language edition published 1985 by
Barron's Educational Series, Inc.

© Parramón Ediciones, S.A.
First Edition, September 1984
The title of the Spanish edition is *el fuego*.

All inquiries should be addressed to:
Barron's Educational Series, Inc.
250 Wireless Boulevard
Hauppauge, New York 11788

ISBN 0-8120-5743-0 (hardcover)
ISBN 0-8120-3598-4 (pbk.)

234   996   12 11 10 9 8

Printed in Spain

the four elements
fire

María Rius
J. M. Parramón

BARRON'S

It's red, very red...

...and orange and yellow...

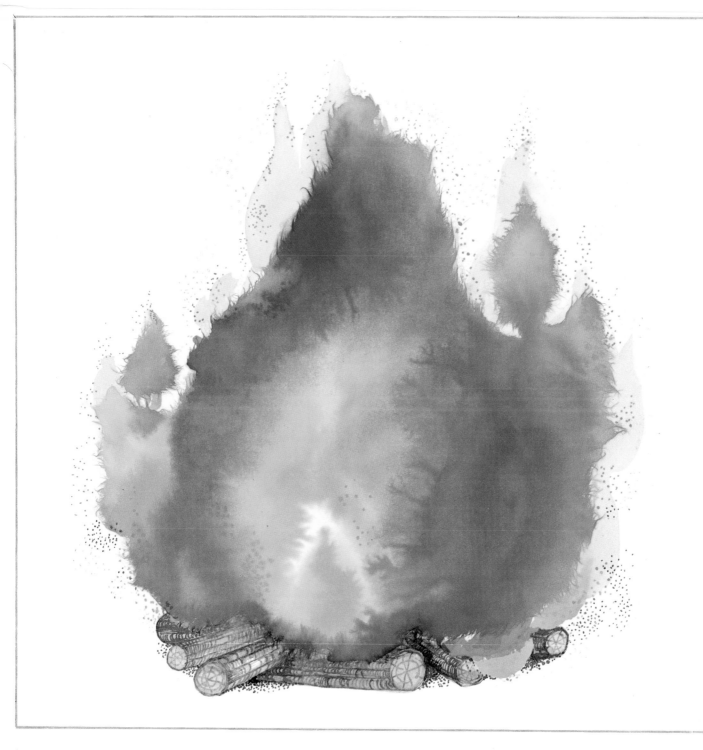

...and blue and white and black

...and always different!

Small, when it's lit...

Big, when it grows...

Terrible, when it spreads...

And it's bad, when it burns...

...but good, when it warms.

It's good for cooking...

and good for launching rockets...

It's good for baking, melting,...creating!

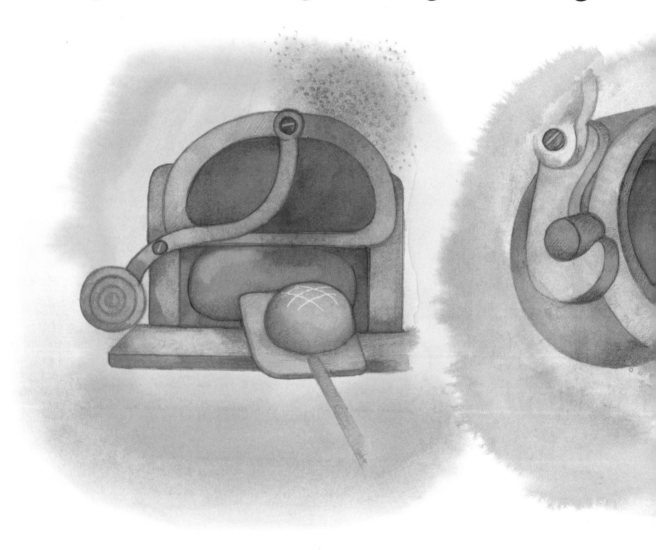

IT'S FIRE!

FIRE

Long live the red flame of fire!
Long live its red flame!

Fire, fire, friend fire,
raise your flame to the wind.
Fire, good friend fire,
Companions in time.

Long live the warmth of fire!
Long live its warmth!

The fire that warms us

How many times, sitting around the fire, have we sung, laughed, felt united. Its light, its colors, its shapes have excited our imagination, and its warmth has sheltered us on long winter nights.

What is heat?

In the olden days people believed that heat was something material, a mysterious fluid that they called *caloric.* If it were material, it had to have weight, and you know that a body weighs the same when it's hot as when it's cold.

If you mix a glass of cold water with one of hot water you get tepid water. What happens? The hot water gives heat to the cold water until they become equal in temperature. We say that heat is "energy in transit" because it shows itself when two bodies that have different temperatures come into contact.

Try to heat an ice cube in a container. The heat that furnishes the fire of your stove is capable of making the ice change state and be converted into liquid water. If you continue heating, the liquid in the container will diminish and you will see steam come out. It is being converted into water vapor! Now its state is gaseous and it stays in the air. Materials can generally present themselves in three states. By giving heat to a solid body, it can pass to a liquid or gas, or the reverse can happen if it loses heat.

Imagine yourself on the tracks of a train. Each piece is separated from the next one by only a fraction of an inch. It is built like this so that it doesn't break when the iron that makes it up expands because of the summer heat. Look at the mercury thermometer. What is its function based on? The majority of substances expand upon heating and contract upon cooling, although they don't all do it in the same proportion.

When we are cooking, why do we use metal spoons with wooden handles? All substances do not conduct heat equally. Metals are *good conductors,* wood is a *bad conductor* (insulator). If the spoon were made entirely of metal, it would burn our fingers.

It's burning

You've seen a candle, the charcoal in a barbecue, a blazing fire…. A substance burns when it combines with oxygen in the air, usually giving out light, heat, and sometimes color. This process is called *combustion.*

They say that substances that can burn are *combustible*. Gasoline is combustible. It burns inside the car's engine. The energy given out in this combustion is used to move the car.

Throw a can into the fire. Metals don't burn easily. If a piece of iron heats up a great deal in a special oven, it becomes "red hot." This state of incandescence is also identified with the word fire.

What is fire?

Fire is ignited matter, with or without the flame, together with light and heat that it gives out.

Practically all the matter that makes up stars is ignited matter. This state, different from solid, liquid, or gas, is considered today as a fourth state of matter, called plasma; it is the most abundant state of matter in the universe.

Humans and fire

The first humans fought to obtain the secrets of fire. Whoever had it was the most powerful. People made bonfires with the sparks of a flint and protected and controlled their fire, using it to cook or roast their food and to defend themselves against certain animals' attacks.

Today fire is in our reach. We know how to use it. *Geothermal* energy uses the internal heat of the earth. Cold water enters an internal high temperature zone. There it is transformed into vapor that in rising will move a turbine connected to a generator, making electricity. *Solar* energy uses the rays of the sun as a source of energy for heating hot water...or for making electricity. This transformation is carried out through solar cells.

The dangers of fire

How is it possible that a dry forest can be in contact with the oxygen in the air, without anything happening, and all it takes is a cigarette butt poorly put out to make everything go up in flames? The embers of the cigarette butt transfer heat energy to the dry leaf it falls on, and the leaf begins to burn. The energy given out in the combustion of the leaf activates the nearby leaves: it gives them a little push so that they, too, begin to burn and so on, successively, until the fire spreads in all directions (unless the wind blows in a certain direction).

Every summer, in hot and dry countries, there are terrible fires. A forest that took centuries to grow is reduced to ashes in a few moments. The forest is dying! And not only the forest, but also its plants and many animals. A few manage to escape, but their home has been destroyed.

Oh fire, you liven our hearts, don't go out with the passing of time, at least leave the ashes of that flame that shines on our youthful years.